Food-Energy-Water Nexus 2019

New York, New York, USA
5 – 6 December 2019

ISBN: 978-1-7138-0579-3

-Printed from e-media with permission by:

Curran Associates, Inc.
57 Morehouse Lane
Red Hook, NY 12571

AIChE
120 Wall Street, FL 23
New York, NY 10005-4020

Phone: (800) 242-4363
Fax: (203) 775-5177

www.aiche.org

Additional copies of this publication are available from:

Curran Associates, Inc.
57 Morehouse Lane
Red Hook, NY 12571 USA
Phone: 845-758-0400
Fax: 845-758-2633
Email: curran@proceedings.com
Web: www.proceedings.com

Food-Energy-Water Nexus 2019

New York, New York, USA
5 – 6 December 2019

AIChE
The Global Home of Chemical Engineers

TABLE OF CONTENTS

PLENARY TALK II

SESSION 1: FEW CASE STUDY AND DECISION MAKING TOOLS

SESSION 2: FOOD WASTE

LIGHTNING ROUND 2

SESSION 3: BIOGAS AND INFRASTRUCTURE

POSTER SESSION

Plenary Talk: OneNYC 2050

- Conference: Food-Energy-Water Nexus
- Year: 2019
- Proceeding: Food-Energy-Water Nexus
- Group: Abstract Submissions
- Session:
  Thursday Invited Talks
- Time:
  Thursday, December 5, 2019 - 9:25am-9:50am

Authors:
Lee, J., NYC Mayor's Office of Sustainability

New York City's Green New Deal is a long-term strategy to secure our city's future. The ambitions of the plan to strengthen our democracy and achieve equity as we confront our climate crisis are grounded in empirical understandings of the impacts of the built environment and the activities of our citizens.

Invited Talk: Catalyzing Commercialization & Societal Impact at NSF

- Conference: Food-Energy-Water Nexus
- Year: 2019
- Proceeding: Food-Energy-Water Nexus
- Group: Abstract Submissions
- Session:
  Thursday Invited Talks
- Time:
  Thursday, December 5, 2019 - 9:50am-10:15am

Authors:
Mehta, R., NSF

NSF's Industrial Innovation and Partnership Division in the Engineering Directorate has number of grant programs to support translational R &D on highly innovative and technically risky projects. This presentation will provide a brief overview of these programs. The reader would benefit from previewing a recent article in this context:
https://www.aiche.org/resources/publications/cep/2019/september/nsf-funding-planting-seeds-innovation
Abstract:
P590829.pdf

Keynote Talk: Responding to the Food-Energy-Water Nexus Challenge

- Conference: Food-Energy-Water Nexus
- Year: 2019
- Proceeding: Food-Energy-Water Nexus
- Group: Abstract Submissions
- Session:
  Thursday Invited Talks
- Time:
  Thursday, December 5, 2019 - 10:15am-10:40am

Authors:
Keairns, D., Deloitte Consulting LLP

The Food-Energy-Water (FEW) Nexus of resources is directly linked to the UN Sustainability Development Goals. The interdependence of the FEW Nexus is a complex, dynamic, engineering, and social challenge. Disruptions, technology innovation, climate change, and social change are common international factors that require consideration in responding to the Nexus challenges. While there are significant differences in each nation or community, there are common themes to incorporate in an approach to addressing the challenges. An international perspective is presented.

Topics:
Food-Energy-Water Nexus

Keynote Talk: FOOD-Energy-Water and Waste Nexus in Achieving Sustainable Circular Urbandevelopment (SCUD)

- Conference: Food-Energy-Water Nexus
- Year: 2019
- Proceeding: Food-Energy-Water Nexus
- Group: Abstract Submissions
- Session:
  Thursday Invited Talks
- Time:
  Thursday, December 5, 2019 - 10:40am-11:05am

Authors:
Guran, S., Rutgers EcoComplex

Food- Energy and Water Nexus is a complex system that requires additional understanding of the ways in which these three pillars function as an integrated system. For instance, there is energy embedded in every gallon of water, there is water embedded in every kWh (or joule) of energy used and every mile travelled, and there is water and energy embedded in every calorie of food humans consume.
Currently most urban systems deal with energy, water and food matters as separate issues and do not consider the intra-impacts of these sectors when trying to achieve their daily operations. Additionally, urban systems has to deal with ever increasing waste matters, decaying urban infrastructure i.e. energy, transportation, water, housing, and climate change related problems. 21st Century urban systems have to be economically and environmentally sustainable while providing social justice to the residents. In order to transform current old cities to tomorrow's more efficient urban locations, these locations can be treated as city-labs to create efficient frameworks that include food, energy water, waste and other material flows. FEW nexus can serve as a tool to achieve Sustainable Circular Urban Development (SCUD).

Topics:
Food-Energy-Water Nexus
Sustainable Engineering

Panel Talk: Building Transdisciplinary, Multi-Stakeholder Partnerships to Facilitate Transitions to Equitable and Sustainable Urban Food Systems

- Conference: Food-Energy-Water Nexus
- Year: 2019
- Proceeding: Food-Energy-Water Nexus
- Group: Abstract Submissions
- Session:
  Panel 1: City-as-lab INFEW RCN: Research Gaps, Current Models and Tools Available for the Stakeholders
- Time:
  Thursday, December 5, 2019 - 10:15am-10:40am

Authors:
Ashton, W., Illinois Institute of Technology Stuart School of Business

US urban food systems are dominated by large corporate forces and are riddled with social, economic and environmental challenges. Socially and economically, low income and minority communities often face disproportionate health burdens as they lack access to healthy, nutritious food, due to low availability and affordability of such food in their neighborhoods. Equally challenging, small producers, particularly minority-owned businesses, in urban and peri-urban areas face challenges with access to capital and formally entering in the food supply chain. Environmental challenges include high energy and water use during production, large carbon footprints in transportation and large volumes of waste throughout the system. Transitions towards equitable and sustainable urban food systems require an in-depth understanding of multi-level and cross sector dynamics in production-consumption chains, including local policies, institutional buying power, local production capacity, and community involvement. Actors across US cities and their surrounding "foodsheds" are developing new and innovative production, distribution and consumption processes, but they face challenges in scaling up and thus increasing their contribution to and impact on local and regional food systems. While demand for local food is increasing, policy mechanisms for scaling local food production and bolstering economic growth were established without considering structural inequalities, consequent environmental degradation, and market dynamics.

Topics:
Food-Energy-Water Nexus

Panel Talk: The Virtual Information Fabric Infrastructure (VIFI): Application Cases for Building Performance Simulations

- Conference: Food-Energy-Water Nexus
- Year: 2019
- Proceeding: Food-Energy-Water Nexus
- Group: Abstract Submissions
- Session:
  Panel 1: City-as-lab INFEW RCN: Research Gaps, Current Models and Tools Available for the Stakeholders
- Time:
  Thursday, December 5, 2019 - 10:40am-11:05am

Authors:
Zhu, Y., Louisiana State University

Although contemporary information and computer technologies have enabled researchers, designers, and engineers to collect large amounts of data, current approaches for sharing data remain limited by concerns such as interoperability, shareability, data size, transport costs, and privacy. These concerns often prevent or confound the development of reliable and accurate building performance simulations. In an effort to overcome these concerns, we envisioned a complementary approach, called the Virtual Information Fabric Infrastructure (VIFI) approach, where data owners share distributed, fragmented data in a manner that does not require the movement of raw data and data users utilize these data for analysis by transporting analytic computation to the data. The VIFI approach presents new opportunities to conduct building performance simulations. Through two case studies, we demonstrated a new computational framework based on VIFI to support an open and collaborative cyberinfrastructure for building performance simulations. Such a computational framework represents a systematic view toward building performance simulations, in which essential components of simulations are coherently integrated.

Topics:
Food-Energy-Water Nexus

Panel Talk: Evaluating the Food-Energy-Water Nexus in an Urban District

- Conference: Food-Energy-Water Nexus
- Year: 2019
- Proceeding: Food-Energy-Water Nexus
- Group: Abstract Submissions
- Session:
  Panel 1: City-as-lab INFEW RCN: Research Gaps, Current Models and Tools Available for the Stakeholders
- Time:
  Thursday, December 5, 2019 - 11:05am-11:30am

Authors:
Eicker, U., Concordia University
Schumacher, J., Concordia University
Braun, R., Reutlingen University
Ghandehari, M., New York University

The work examines how urban data and urban models helps to characterize the food, energy and water Nexus on an urban district level. A bottom up approach is used to model urban system development.
To assess the energy-saving potential in the food sector in Borough Hall district, the status quo is analyzed. Open data from OpenStreetMap, NYCOpenData and PLUTO were used to locate the food related usages and to determine the size of each location.
The calculations show, that the electricity demand for the food sector in Borough Hall is 70 GWh/a. With a total electricity consumption of 1.33 TWh/a (2013), 70 GWh/a equals 5.2% of the whole electricity demand of the district of the total consumption (1% for food sales and 4.2% food services). For food refrigeration, almost 32 GWh/a electrical power is used, therefore refrigeration bears the most potential in terms of energy saving.
From all the organic waste in Borough Hall, 14 427 tons annually are digester usable. This amount can be converted into 1 313 000 m³/a of biogas, and then used in a combined heat and power plant to produce 3.9 GWh/a heat and 3.2 GWh/a electricity. Another possibility is to transform the whole biogas quantity into 787 713 m³/a of methane and add it into the gas grid. Finally, biomethane can work as a fuel and the amount is enough to fuel 1 040 cars for a year.
Topics:
Food (Sustainability & Environment)
Computing and Systems Engineering
Alternative Energy

Panel Talk: Lessons Learned: Creating an Interdisciplinary Team to Address a FEW Nexus Hotspot in San Antonio, Texas

- Conference: Food-Energy-Water Nexus
- Year: 2019
- Proceeding: Food-Energy-Water Nexus
- Group: Abstract Submissions
- Session:
  Panel 1: City-as-lab INFEW RCN: Research Gaps, Current Models and Tools Available for the Stakeholders
- Time:
  Thursday, December 5, 2019 - 11:30am-11:55am

Authors:
Daher, B., Texas A&M University

San Antonio is home to a rapidly urbanizing population, with major agricultural activity surrounding the city, and a growing production of oil and natural gas in its underlying Eagle Ford shale play. The region of San Antonio represents a resource hotspot whose stakeholders compete across sectors for the same limited water, land, and financial resources and whose projection trends indicate continued growth across those sectors. This presentation will include lessons learned from the experience of creating a campus wide interdisciplinary group to address the resource challenges facing the San Antonio region at Texas A&M's Water-Energy-Food Nexus Initiative (WEFNI). The presentation will also present a brief overview of the questions and research conducted under thematic foci that include data and modeling, trade-off analysis, water for food, water for energy, and governance.
Topics:
Food-Energy-Water Nexus

Panel Talk: Natural Language Processing and Spatiotemporal Visualization for Wildfire Event Reconstruction

- Conference: Food-Energy-Water Nexus
- Year: 2019
- Proceeding: Food-Energy-Water Nexus
- Group: Abstract Submissions
- Session:
  Panel 1: City-as-lab INFEW RCN: Research Gaps, Current Models and Tools Available for the Stakeholders
- Time:
  Thursday, December 5, 2019 - 11:55am-12:20pm

Authors:
Flaxman, M., Geodesign Technologies

Fire management is at a crisis point. A century of fire suppression, extensive urban wildland interface development, and climate change have led to nearly unstoppable fires. Attempts to avoid some of these fires using public power supply shutoffs have adversely affected tens of millions of citizens and cost billions of dollars. Major wildfire events have for many decades been documented with "after action reports." More than 3,000 such reports are publicly available. These describe the chronology of events, including weather, fire behavior, people and equipment deployed. The impacts of each fire on lives, properties and the environmental are carefully documented. Historically, such reports have included only very limited spatial data. This makes it very difficult to comprehend narrative references to fuels, topography and local landmarks. But elsewhere within State and Federal databases, extensive spatial data are available, detailing environmental condition before, during and after both prescribed fires and wildfires. Our project is using a Natural Language Processing approach to extract relevant entities from free text, automatically linking them to spatial data, and using the combined datasets to parameterized historical fire scenarios within standard fire simulation models. We use a machine learning model (SpaCy BERT) to perform "named entity recognition," including the identification of weather, people, equipment, institutions and geographic places as objects. Geographic places are cross referenced to fire names, dates and states within GeoMac and MTBS fire perimeter databases. These perimeters are in turn used to extract information from Landfire and analysis-ready satellite data archives suitable for fire modeling.

Topics:
Food-Energy-Water Nexus

On the Role of Energy Storage in Future Energy Networks

- Conference: Food-Energy-Water Nexus
- Year: 2019
- Proceeding: Food-Energy-Water Nexus
- Group: Abstract Submissions
- Session:
  Panel 1: City-as-lab INFEW RCN: Research Gaps, Current Models and Tools Available for the Stakeholders
- Time:
  Thursday, December 5, 2019 - 12:20pm-12:45pm

Authors:
Mohamed, A., City College of the City University of New York

Urban systems, including food, water, and transportation, heavily depend on the energy network for energy supply. Energy Storage Systems (ESS) are poised to radically change the way the energy network, and consequently other urban systems, are designed and operated. This talk will highlight some of the benefits and barriers of ESS, and will provide some real-world NYC-specific examples of ESS applications, initiatives and demonstration projects.

Topics:
Food-Energy-Water Nexus

A Study on Food, Energy, and Water Impacts of the Average American Diet By Demographic Group

- Conference: Food-Energy-Water Nexus
- Year: 2019
- Proceeding: Food-Energy-Water Nexus
- Group: Abstract Submissions
- Session:
  Lightning Round 1
- Time:
  Thursday, December 5, 2019 - 2:20pm-2:25pm

Authors:
Bozeman, J. III, Institute for Environmental Science and Policy, University of Illinois at Chicago

Overcoming socioecological barriers of the food–energy–water (FEW) nexus remains a challenge for urban and rural communities. Using life cycle assessment and social psychology techniques, cradle-to-farm-gate land, greenhouse gas (GHG), and water impacts—that derive from food consumption in the United States—were analyzed and differentiated by major demographic groups (Black, Latinx, and White). Results indicate that the White demographic yields the highest per capita GHG (680 kg of CO_2 eq*year^{-1}) and water impacts (328,600 L*year^{-1}) from food consumption, whereas the Black demographic yields the highest per capita land impacts (1,770 m^2*year^{-1}) from food consumption. Furthermore, we propose using government programming to encourage environmentally friendly food purchasing behavior. In all, these findings establish a cross-disciplinary methodological approach to addressing socioecological problems and provides useful FEW impact data for FEW nexus and climate change researchers.

Abstract:
P588230.pdf
Topics:
Food-Energy-Water Nexus
Environmental Sustainability

Is Organic Still Sustainable? an Investigation of Consumer Perceptions of Food Production Sustainability

- Conference: Food-Energy-Water Nexus
- Year: 2019
- Proceeding: Food-Energy-Water Nexus
- Group: Abstract Submissions
- Session:
  Lightning Round 1
- Time:
  Thursday, December 5, 2019 - 2:25pm-2:30pm

Authors:
Cho, C., New York Institute of Technology
Chen, S., New York Institute of Technology
DaSilva, J., New York Institute of Technology
Flores, J., New York Institute of Technology
Sotiropoulos, V., New York Institute of Technology

There is a belief that organic food is sustainable. The problem that has arisen is that organic food may no longer be as sustainable as it once was. Initially, organic food was produced by local and small production farms. Today, however, they are mass-produced and may be engaging in the same kind of carbon-producing practices as mass-produced goods. Nonetheless, despite the change in production, many consumers continue to believe that organic foods are more sustainable than their nonorganic counterparts. This may be due to the curated image of organic food through marketing. The local community displays organic food as sustainable because it is natural, yet it is unclear if organic food actually contains any growth promotion supplements. The problem then becomes that the consumption of organic foods will increase; causing the same kind of damage to the environment as nonorganic foods. More research is needed to uncover this potentially harmful production practice as more organic food is purchased. The goal of this project is to assess consumer misperceptions related to the sustainability of organic foods versus nonorganic foods. The expectation here is consumers will perceive organic foods to be far more sustainable than they actually are. Using a survey-based approach, a sample of 80 participants will be asked about their perceptions of food sustainability practices (i.e., cost, carbon footprint emission, water electricity consumption) of organic versus nonorganic foods. Results, findings, implications, limitations and future limitations will also be discussed.

Topics:
Sustainability & Environment

A Case Study of Dietary Pattern Change Impact on Carbon Footprint

- Conference: Food-Energy-Water Nexus
- Year: 2019
- Proceeding: Food-Energy-Water Nexus
- Group: Abstract Submissions
- Session:
  Lightning Round 1
- Time:
  Thursday, December 5, 2019 - 2:30pm-2:35pm

Authors:
Elangovan, R., New York Institute of Technology
Dong, Z., New York Institute of Technology

A significant portion of greenhouse gas emissions have been attributed to the food sector, but little is known about the association between the carbon footprint of individual diets and types of food consumed. Development and expansion of agriculture are essentially linked to dietary choices and contribute to deforestation, degradation of land, biodiversity loss, extensive freshwater use, and water pollution. We present a case study of the impact of dietary patterns on carbon footprint at the Gowanus district in the Borough of Brooklyn in NYC. The study considers the population projection within this district due to rezoning and assess the impact of a dietary pattern change in the carbon footprint and to understand their association with different gender and age groups. The carbon footprint examined for every food item includes the emission from all the processes of a food life cycle. A recent US Dietary Guidelines Advisory Committee claimed that the food patterns responsible for lower greenhouse gas emission has a better overall diet quality and are healthier on several key dimensions. Based on our study, we inferred that the consumption of meat contributes to almost 90% of the total emission for the current dietary pattern. A diverse dietary pattern can have a significantly impact on the food carbon footprint. We show a 50% reduction in meat consumption may reduce carbon emission by 45%.

Abstract:
P588361.pdf
Topics:
Food-Energy-Water Nexus

Utilizing Urban Organic Wastes Via Hydrothermal Carbonization Process for Energy Production

- Conference: Food-Energy-Water Nexus
- Year: 2019
- Proceeding: Food-Energy-Water Nexus
- Group: Abstract Submissions
- Session:
  Lightning Round 1
- Time:
  Thursday, December 5, 2019 - 2:40pm-2:45pm

Authors:
Satrio, J. A., Villanova University
Lee, R., Villanova University
Taylor, J., Gray Brothers - Somax Envirinmental
Spracklin, D., Gray Brothers- Somax Environmental Company

Organic waste is a major component of landfills, which is the largest source of methane emissions in the world. In the USA, on food waste alone, from harvesting to consumption, excluding the non-edible parts, 30-40% of the food is wasted, equaling more than 35 million tons per year or approximately 240 pounds of food per year per person. Finding ways to utilize this wet organic waste into useful products will be desirable from both economic and environmental sustainability points of view.

Villanova University and Gray Brothers-Somax Environmental Company are working together in developing thermochemical-based process, namely hydrothermal carbonization (HTC), for converting wet organic wastes in urban communities, specifically food wastes and sewage sludge, into coal-like product called hydrocar. Under sub-critical conditions, water in the HTC reactor stays in liquid form and acts as a reaction medium to promote the breakdown and cleavage of chemical bonds in the organic solids. The absence of liquid-to-vapor phase change of the water makes the process significantly less energy intensive compared to a process that involves water vaporization or drying, such as incineration or gasification. The temperature range for HTC is from 180-240°C producing solid hydrochar as the main product.

The focus of the present study is to evaluate the effects of feedstock and HTC process parameters on the yields and physiochemical properties of hydrocar products and obtain the optimal process conditions which will used for scale-up design of the HTC reactor system.

Topics:
Alternative Energy
Biofuels (Energy)

Machine Learning Approach to Examine the Relationship between Residential Energy and Water Consumption with Greenhouse Gas Emissions in NYC

- Conference: Food-Energy-Water Nexus
- Year: 2019
- Proceeding: Food-Energy-Water Nexus
- Group: Abstract Submissions
- Session:
  Lightning Round 1
- Time:
  Thursday, December 5, 2019 - 2:45pm-2:50pm

Authors:
Brown, S., New York Institute of Technology
Baidas, M., New York Institute of Technology
Dong, Z., New York Institute of Technology

The City of New York has set an ambitious goal to reduce 80% of its greenhouse gas (GHG) emission by 2050, relative to the 2005 baseline. While approaches have been taken on the regulatory level, residential consumption of water and energy needs to be better understood for incentives on reducing house-hold level GHG emissions. This study examines the impact of end-use residential water and energy consumption on greenhouse gas emissions. We model the relationships between demand, end-use consumption, and emissions. We use artificial neural networks to predict the impact of the water-energy nexus on greenhouse gas emissions. We also examine how these relationships change by the income of the area of consumption. We theorize that income impacts residential energy and water consumption and consequently, GHG emissions outcomes.

Topics:
Food-Energy-Water Nexus

Simultaneous Ammonia-Phosphate Recovery from Wastewater As Struvite Fertilizer for Food Quality

- Conference: Food-Energy-Water Nexus
- Year: 2019
- Proceeding: Food-Energy-Water Nexus
- Group: Abstract Submissions
- Session:
  Lightning Round 1
- Time:
  Thursday, December 5, 2019 - 2:50pm-2:55pm

Authors:
Camacho, L. M., Texas A&M University-Kingsville
Ullah, M. A., Texas A&M University-Kingsville

Rapid world population growth, climate change, increasing urbanization, and excessive nutrient inputs into open bodies of water are pressing on wastewater utilities to provide treated water in an energy-efficient manner that promotes human health and protection to the environment. Expected increased population will exert a huge demand for food, thus scarcity of nutrient requirements in food production will prevail. Wastewater contains a high amount of organic matter, nitrogen and phosphorus, and considerable amounts of magnesium, which are the potential source for nutrient recovery. In this study, ammonia and phosphorus were simultaneously recovered from a wastewater effluent by using natural and modified Clinoptilolite zeolites. The adsorbent materials were tested in a semi-pilot system using simulated and real wastewater. The recovered nutrients were efficiently converted into Struvite, a fertilizer that can help to save food, energy, and water. The used zeolites were recovered and reuse to recover additional nutrients.

Topics:
Water (Sustainability & Environment)
Energy (Sustainability & Environment)
Food-Energy-Water Nexus

CO2-Responsive Tunable Phase Changing Nano-Scale Fluids for Subsurface Energy Recovery

- Conference: Food-Energy-Water Nexus
- Year: 2019
- Proceeding: Food-Energy-Water Nexus
- Group: Abstract Submissions
- Session:
  Lightning Round 1
- Time:
  Thursday, December 5, 2019 - 2:55pm-3:00pm

Authors:
Asgar, H., Cornell University
Gadikota, G., Cornell University
Huang, E., Cornell University

Rising consumption of water for subsurface energy recovery has motivated the exploration of novel fluids to enhance the permeability of the subsurface geologic reservoirs. Advancements in developing novel nanofluids will allow us to enhance heat recovery from geothermal reservoirs and reduce water use for hydraulic fracturing operations. Specifically, we explore the design of CO_2-responsive phase changing nanofluids. These nanofluids are constructed from SiO_2 nanoparticles tethered with poly(allylamine) (PAA), which form hydrogel upon reaction with CO_2. The presence of SiO_2 nanoparticles enhances the CO_2 absorption capacity when compared with the pure PAA solution. The CO_2 absorption on PAA and SiO_2-PAA was governed by the formation of carbamate, protonated primary and secondary, and bicarbonate ions using infrared spectroscopy. Time resolved USAXS/SAXS measurements showed an increase in the sizes of SiO_2-PAA aggregates during CO_2 loading and hydrogel formation. Additionally, the effect of nanoparticle size and concentration of polymeric chains on CO_2 absorption was also investigated. These observations point to the feasibility of enhancing hydrogel formation with these novel nano-scale fluids to enhance permeability in the subsurface environment.

Topics:
Energy (Sustainability & Environment)
Environmental Sustainability

Upcoming Scenarios and Developing Trends in the Generation of Energy in Pakistan: Forecasting Supply and Demand

- Conference: Food-Energy-Water Nexus
- Year: 2019
- Proceeding: Food-Energy-Water Nexus
- Group: Abstract Submissions
- Session:
  Lightning Round 1
- Time:
  Thursday, December 5, 2019 - 3:00pm-3:05pm

Authors:
Danish, M., University of Engineering and Technology, Lahore
Farooq, U., East China University of Science and Technology, Shanghai
Ahmad, S., University of Engineering and Technology, Lahore
Ahmad, S. W., University of Engineering and Technology, Lahore

The energy matrix structure of Pakistan shows least generation of power from renewable sources. The contribution of renewable sources for the production of energy in 2019 is just 5-6%. An exceptional potential is existing for the expansion of the renewable energy production in Pakistan that can even outweighs the national demands of the energy. An outstanding contribution of hydropower is existing currently in the national energy matrix. Pakistan has a significant potential for the exposure of other renewable energy sources such as solar, biomass and wind. This study focuses on the evolution of the Pakistani energy matrix exposing probable mitigation scenarios keeping in view of the possible climate changes. The LEAP System (Long range energy alternatives planning) methodology was adopted for the modeling purpose that allows us to suggest diverse scenarios under the umbrella of socio-economic scenarios and the REGSA developed project (Promoting renewable electricity generation in South America). The evaluated results forecast the future scenarios, probable developments in energy generation in terms of projected demand and supply till 2040 in Pakistan.
Topics:
Energy Policy

Plenary Talk: A Global Crisis - Beyond The Brink

- Conference: Food-Energy-Water Nexus
- Year: 2019
- Proceeding: Food-Energy-Water Nexus
- Group: Abstract Submissions
- Session:
  Plenary Talk II
- Time:
  Friday, December 6, 2019 - 9:00am-9:45am

Authors:
Thebaut, J., Chronicles Group

The U.S. National Intelligence Council's and other authoritative voices predict a dangerous future for the world, including unprecedented consequences of warming temperatures, explosive population growth, massive water and food shortages, global pollution, widespread human migration, armed conflict involving national and subnational groups, and the mounting threat of nuclear war. Human civilization is at a crossroads to survive; and the challenge is to establish world-wide educational initiative that will energize citizens to take positive action. Chronicles Group, Inc., a 501 (c)(3) corporation, proposes to create a 13-part, 30-minute episodic television series and a theatrical documentary feature that will first, place a spotlight on the evolving global crisis, and second, highlight immediate solutions for the growing food, energy and water humanitarian crises. In order to maintain the highest journalistic integrity, the Chronicles Group will conduct on-camera interviews with distinguished scientists, experts, and solutions-focused stakeholders. The production will be designed to be entertaining, compelling, provocative, and accurate. The 2017 Nobel Peace Prize recipient Ira Helfand stated in a recent article that America confronts a long list of critical problems that all require urgent attention. Among them, two drivers of global climate disruption stand out. Fossil-fuel use and nuclear war are unique in the threat they pose to global climate, and thus to the very survival of human civilization. In recent remarks UN Secretary-General Antonio Guterres' called climate change the defining issue of our time, with now being a defining moment...complete abstract in program book.

Abstract:
P589538.pdf
Topics:
Food-Energy-Water Nexus

Keynote Talk: BioEnergy with Carbon Capture and Storage (BECCS): H2 production from food wastes with suppressed COx formation

- Conference: Food-Energy-Water Nexus
- Year: 2019
- Proceeding: Food-Energy-Water Nexus
- Group: Abstract Submissions
- Session:
  Session 1: FEW Case Study and Decision Making Tools
- Time:
  Friday, December 6, 2019 - 10:00am-10:24am

Authors:
Park, A. H. A., Columbia University

Biomass is considered a carbon-neutral energy source and can even be carbon-negative if combined with a carbon capture and storage (CCS) scheme. The main challenges with bioenergy include the limited availability of biomass compared to fossil resources and the change in land use leading to net CO_2 emissions. Food wastes and seaweeds offer an advantageous alternative biomass for biofuel production. They are widely available in many parts of the world. Thus, the use of food wastes, seaweed or algae growing in seawater would allow many countries with limited land and fresh water to produce bioenergy. Unfortunately, the bioenergy technologies that can convert wet and salty biomass such as food wastes and seaweed are very limited, representing a significant untapped opportunity for biofuels. The alkaline thermal treatment of biomass provides a unique reaction pathway to produce H_2 while in-situ capturing CO_2. By directly reacting wet and salty biomass with alkaline metal hydroxide, high purity H_2 is generated while capturing and storing CO_2 as solid carbonate.

Topics:
Food-Energy-Water Nexus

How 50 Food & Beverage Manufacturing Facilities Improved Profitability By Reducing Food Loss

- Conference: Food-Energy-Water Nexus
- Year: 2019
- Proceeding: Food-Energy-Water Nexus
- Group: Abstract Submissions
- Session:
  Session 1: FEW Case Study and Decision Making Tools
- Time:
  Friday, December 6, 2019 - 10:30am-10:45am

Authors:
Taylor, B., Enviro-Stewards Inc.

One third of food produced is presently wasted. And if food loss were a country, food loss would be the third largest GHG emitter (after USA & China). However, even if we succeeded at diverting 100% of this food from landfill, one third of food would still be wasted and it would remain the third largest GHG emitter.

Enviro-Stewards and Provision Coalition completed a pilot food loss prevention assessment for Campbell Soup that identified practical measures to prevent $706,000/yr of food loss. This project was selected as the top clean capital project in Canada. Subsequently, the Walmart Foundation provided project funding to The Canadian Centre for Food Integrity to further demonstrate prevention opportunities at 50 food & beverage manufacturing facilities across Canada. The facilities assessed by Enviro-Stewards for this project included bakeries, dairies, breweries, vegetable processors, meat processors, seafood companies, among others.

The 50 assessments averaged $230,000/yr of savings per facility with less than one year payback period. This presentation outlines Enviro-Stewards' project approach used to identify and quantify these savings as well as illustrations (with case studies) of practical measures implemented to secure the economic, social and environmental benefits available through food loss prevention.

Abstract:
P584128.pdf
Topics:
Sustainable Engineering
Plant Operations
Food-Energy-Water Nexus

Toward Creating an Environment of Cooperation between Water, Energy,and Food Stakeholders in San Antonio

- Conference: Food-Energy-Water Nexus
- Year: 2019
- Proceeding: Food-Energy-Water Nexus
- Group: Abstract Submissions
- Session:
  Session 1: FEW Case Study and Decision Making Tools
- Time:
  Friday, December 6, 2019 - 10:42am-11:00am

Authors:
Daher, B., Texas A&M University
Mohtar, R., American University of Beirut
Portney, K., The Bush School of Government and Public Service, Texas A&M University
Hannibal, B., Texas A&M University

The San Antonio Region is home to a rapidly growing population with developing energy and agricultural sectors competing for water, land, and financial resources. Despite the tight interconnectedness between water, energy, and food challenges, little is known about the levels of communication and coordination among the various officials responsible for making the decisions that affect the management and planning of the three resource systems. It has been postulated that efficient communication is a prerequisite to developing resource allocation strategies that avoid potential unintended negative consequences that could result from inefficient allocation of natural resources and competing demands. Factors that may impact communication are identified and their potential roles are considered in improving existing levels of communication between San Antonio's water officials and those at other energy, food, and water institutions in the San Antonio Region. A questionnaire designed to gather information on stakeholder concerns, frequency of communication, and participation in engagement forums was sent to public water officials in the Region. Using social network analysis and bivariate Ordinary Least Square regression analysis, the authors conclude that while modest levels of communication exist among water institutions, a very low level of communication exists between water institutions and those responsible for food and energy. Improving this level of communication could be partly be achieved through investing in cross-institutional mechanisms which promote higher levels of cooperation, and that work toward improving the compatibility of differing planning horizons, and common goal setting activities across sectors.

Topics:
Food-Energy-Water Nexus

Integrated Modeling of Food-Energy-Water Systems: Challenges and Opportunities of Quantitative Graphical Networks

- Conference: Food-Energy-Water Nexus
- Year: 2019
- Proceeding: 2019 Food-Energy-Water Nexus
- Group: Abstract Submissions
- Session:
  Session 1: FEW Case Study and Decision Making Tools
- Time:
  Friday, December 6, 2019 - 11:00am-11:18am

Authors:
Calder, R. S. D., Duke University
Jeuland, M., Duke University
Borsuk, M. E., Duke University
Olander, L. P., Duke University
Bradbury, K., Duke University
Malof, J. M., Duke University

Food, energy and water systems are under increasing stresses from population growth, development and depletion of natural resources. There is ample modeling capacity for energy-food, water-food and energy-water relationships, but each of these pairwise relationships makes strong assumptions about the others and neglects the role of interacting social and environmental variables. Consequently, interventions in food, energy and water systems have tended to have unpredictable and unintended outcomes. Our work explores the capacity of graphical network approaches to characterize tightly coupled human-natural systems, resolve paradoxes and better anticipate the impacts of policies and investments. We aim to then leverage these approaches into quantitative decision-support tools. We find that evidence synthesis and interpretation can be improved by decomposing high-level associations of interest (e.g., the relationship between crop yield and investments in irrigation) into a graphical structure that accounts for interacting socio-environmental variables. However, the ability to parameterize such detailed models is highly influenced by availability, quality and scale of data. We hypothesize that emerging methods in remote sensing and machine learning can be deployed to address these needs and that the resulting models will improve decision-making in the realm of food, energy and water systems.

Topics:
Water (Sustainability & Environment)
Food-Energy-Water Nexus

Investigating the Food-Energy-Water Nexus through Embedded Resource Accounting

- Conference: Food-Energy-Water Nexus
- Year: 2019
- Proceeding: Food-Energy-Water Nexus
- Group: Abstract Submissions
- Session:
  Session 1: FEW Case Study and Decision Making Tools
- Time:
  Friday, December 6, 2019 - 11:18am-11:36am

Authors:
Mahjabin, T., The Pennsylvania State University
Grady, C., Pennsylvania State University

Food-energy-water (FEW) nexus frameworks can enhance our understanding of interconnections, feedback, and dependencies in the FEW systems. Complex, connecting stressors such as increasing population, urbanization, climate change, and economic development will all influence the sustainable usage of food, energy, and water. This presentation will highlight a variety of techniques from the field of embedded resource accounting, including water footprint and nitrogen footprint tools, to build knowledge around the FEW nexus. These embedded resource estimation or footprint methods allow for a comparison between water, food, and energy under one unit of measurement. To investigate the water footprint efficiency across the US, this research examines the scaling properties of water footprint production and consumption with various sociodemographic information. Using network science approaches, we also investigate the structure and characteristics of the virtual water networks embedded in US food and energy transfers. This study also develops an interstate network of virtual nitrogen (N) transfer embedded in food production and compares network characteristics of the virtual N to the virtual water transfer network. Finally, the study examines uncertainty in the footprint values due to the variability in the commodity trade across the US. These research pursuits provide tools and techniques using open-source information to inform understandings across the FEW nexus.
Abstract:
P588370.pdf
Topics:
Water (Sustainability & Environment)
Food (Sustainability & Environment)

Wrf's Transition from Water-Energy Nexus to the Food-Energy-Water Nexus

- Conference: Food-Energy-Water Nexus
- Year: 2019
- Proceeding: Food-Energy-Water Nexus
- Group: Abstract Submissions
- Session:
  Session 1: FEW Case Study and Decision Making Tools
- Time:
  Friday, December 6, 2019 - 11:36am-11:54am

Authors:
Dhanasekar, A., The Water Research Foundation

Energy has always been an important aspect as far as water is concerned. The Water Research Foundation (WRF) has had a research area solely focused on energy efficiency and resource recovery for more than a decade now. With the increasing availability of efficient and economic technologies, now is the time to develop new ways to increase energy efficiency at water and wastewater utilities as well as optimizing the use of energy for operations. WRF's portfolio of energy research has increased considerably to include all different methods of solving issues related to energy as a puzzle. Initially, the research was focused on providing effective strategies to reduce water utility energy consumption and cost, developing strategies for multi-sector, regional, integrated water-energy planning, and providing sound approaches for energy generation by water utilities. Resource recovery is now more important than ever, and the implementation of distributed energy resources and renewables is more convenient and possible with more supportive policy and incentives that help economic implementation. This presentation will highlight the direction of energy research that WRF is heading towards which includes increased sustainability, optimized energy management, big data in water and wastewater utilities and innovative technologies to enhance resource recovery, leading into the Food-Energy-Water Nexus. The primary objective of this presentation will be to showcase the latest trends in Energy Research and how WRF is leading the efforts with an overview of some its latest research findings.

Topics:
Food-Energy-Water Nexus
Energy Efficiency
Water (Sustainability & Environment)

The Intersection of Urban and Rural Communities: Building Resilience through the Food Energy Water Nexus

- Conference: Food-Energy-Water Nexus
- Year: 2019
- Proceeding: Food-Energy-Water Nexus
- Group: Abstract Submissions
- Session:
  Session 1: FEW Case Study and Decision Making Tools
- Time:
  Friday, December 6, 2019 - 11:54am-12:12pm

Authors:
Daw, J., National Renewable Energy Laboratory
Stout, S., National Renewable Energy Laboratory

This presentation will highlight integrated strategies and policies to build resilience across energy, water and food systems using case studies from NREL's work in islands, remote communities, urban, and rural areas to highlight the intersection and opportunity for improved urban-rural system resilience. While urban resilience strategies are not necessarily applicable to rural settings, they are inherently interrelated. The majority of America's agricultural production occurs in rural areas, and urban centers are dependent on energy, water, and food systems that exist and intersect in rural areas outside of their jurisdictional boundaries.

Key differences between urban and rural contexts, including distinct built environments (e.g., building types, density, transportation networks, energy and water infrastructure), economic drivers, and other physical and demographic factors highlight the need for integrated planning of urban and rural energy-water-food systems. These strategies are important for rural communities themselves, and the urban populations that depend on rural areas for energy, water, and food services.

Extreme events have highlighted weaknesses in energy, water, and food systems as a result of supply chain disruptions and cascading system failures. These events have challenged both urban and rural communities to adopt policies that improve their resilience and to seek new strategies to ensure continued security of energy, water, and food resources. Continued access to energy, water and food is not just essential in post disaster scenarios, it also fuels the socioeconomic growth and prosperity of rural communities, where the mindful management of energy, water and food systems is increasingly critical to their survival.

Topics:
Food-Energy-Water Nexus
Energy (Sustainability & Environment)
Water (Sustainability & Environment)

Keynote Talk: Food Waste As a Sustainable Resource for the Production of Carbons for Purification and Catalytic Processes

- Conference: Food-Energy-Water Nexus
- Year: 2019
- Proceeding: Food-Energy-Water Nexus
- Group: Abstract Submissions
- Session:
  Session 2: Food Waste
- Time:
  Friday, December 6, 2019 - 1:00pm-1:24pm

Authors:
Valla, J. A., University of Connecticut
Yu, L., University of Connecticut
Gamliel, D. P., Physical Sciences Inc.
Markunas, B., University of Connecticut
Saltzgiver, K. M., University of Connecticut

Food waste can be considered as a renewable resource for energy and/or materials production. In this study, biochars were prepared via pyrolysis of food waste and then used as precursors for production of activated carbons via physical and chemical activation. Pyrolysis and activation conditions were varied to optimize the properties of biochar and activated carbons, respectively. Activated carbons were characterized by N_2 sorption-desorption, scanning electron microscope (SEM), energy-dispersive X-ray spectroscopy (EDX), elemental analysis, inductively coupled plasma atomic emission spectroscopy (ICP-OES), X-ray powder diffraction (XRD), ^{13}C nuclear magnetic resonance (NMR), Fourier-transform infrared spectroscopy (FTIR) and Raman spectroscopy. Adsorption experiments of aromatic hydrocarbons were conducted to evaluate the feasibility of using biochars and activated carbons derived from food waste for the adsorption of aromatic hydrocarbons (benzene and naphthalene) from water. The activated carbon with the highest adsorption capacity was prepared via steam activation at 950ºC/1 h. The BET surface area, micropore and total pore volume of the activated carbon were 745 m^2/g, 0.185 cm^3/g and 0.594 cm^3/g, respectively. The benzene and naphthalene sorption capacity was 460 mg/g and 150 mg/g, respectively. The physicochemical properties and adsorption capacities of the activated carbons prepared from food waste were also compared to the corresponding properties and capacities of the activated carbons derived from miscanthus, as well as commercial carbons. Our study revealed that activated carbons prepared by physical activation of biochars derived from food waste are promising candidates for water purification with capacities for aromatic hydrocarbons similar to the ones observed using commercial activated carbons.

Topics:
Food (Chemicals & Materials)
Separations

Topic: Food Waste in K-12 Schoolstitle: Food Waste Prevention and Reduction Pilot in Baltimore City Schools: Food Matters - for Students, Faculty and Staff

- Conference: Food-Energy-Water Nexus
- Year: 2019
- Proceeding: Food-Energy-Water Nexus
- Group: Abstract Submissions
- Session:
  Session 2: Food Waste
- Time:
  Friday, December 6, 2019 - 1:24pm-1:42pm

Authors:
Richardson, A., Baltimore City Government

Wasted food has gained much attention in recent years both nationally and globally. Despite this recent focus, excessive food waste has been a burgeoning issue for decades. Food insecurity and hunger - perplexingly - continue concomitantly while landfills are increasingly fed opposed to people. Many organizations are focusing on K-12 schools to help curb rampant food waste as both significant contributors to food waste and vehicles for food waste prevention education among students. To appreciate the totality of resources lost when food is wasted, the authors uses the food-energy-water (FEW) nexus framework to assess the total cost of resources lost when food is wasted. FEW has become part of the food systems lexicon used to describe food activity in the current anthropogenic era. This framework is a useful tool for assessing regional food waste challenges and opportunities. Breaking the nexus down into three components - water/food; food/energy and water/energy, can help local, state and Federal governments achieve more efficient management of natural resources. With an estimated 40% of all food grown in the U.S. going to landfills and incinerators, the management of FEW systems leaves much to be desired. A FEW framing can connect with various curriculum elements from math to science to reading - weaving in FEW constructs can help students gain an appreciation for the environmental impact of our food system at an early age. These learnings can foster a food waste reduction and prevention culture among students that shapes their perspective and behaviors into adolescence and adulthood

Topics:
K-12
Food (Sustainability & Environment)
Food-Energy-Water Nexus

Food, Energy & Water: The Impact of a Food Waste Reduction Training Program for Food Service Workers in an Urban School District

- Conference: Food-Energy-Water Nexus
- Year: 2019
- Proceeding: Food-Energy-Water Nexus
- Group: Abstract Submissions
- Session:
  Session 2: Food Waste
- Time:
  Friday, December 6, 2019 - 1:42pm-2:00pm

Authors:
Elnakib, S., Rutgers Cooperative Extension
Rowe, A., Rutgers Cooperative Extension

Food waste is a major issue in the United States. The United States Department of Agriculture (USDA) estimates that 30-40% of the food produced in the United States is wasted.

15 Paterson, NJ, schools that cook their own meals were randomly chosen to participate in the study, with the intervention based on USDA's offer vs. serve model. Lunch food waste data collection occurred prior to the intervention, then the food service workers and lunchroom monitors were trained. Finally, the post-intervention food waste was measured. Training for the school foodservice workers was based on behavioral economics "nudges" to improve the cafeteria environment which led to reduced food waste.

Of the food and beverages served during our 60 visits to schools 2,473 pounds were wasted before the intervention and 2,123 were wasted after the intervention. Overall, 350 pounds of food was saved which was a 14 % reduction in food waste due to this intervention. That is approximately 12 pounds of waste saved per school per day and a total of 90,720 pounds of food waste saved for the whole district for the year. The estimated savings of food cost for the 90,620 pounds of food is $76,452. Additionally, an estimated 41,932 gallons of water were saved due to the intervention. Overall, the intervention showed the impact of a food service training program on reducing food waste.

Topics:
Food (Sustainability & Environment)
Water (Sustainability & Environment)
Environmental Sustainability

Focus on Food, Farming and Climate Change

- Conference: Food-Energy-Water Nexus
- Year: 2019
- Proceeding: Food-Energy-Water Nexus
- Group: Abstract Submissions
- Session:
  Session 2: Food Waste
- Time:
  Friday, December 6, 2019 - 2:00pm-2:18pm

Authors:
Starr, J., Cargill

In looking at the greenhouse gas emissions associated with the food on your table, much of those emissions are a result of activities on the farm. Given the large number of farmers in the US and worldwide, it is a significant task to change farming practices to more sustainable methods. These more sustainable methods may look more expensive than their current practices to a farmer who is seeing low prices for his current crop. It will take innovation and good science to align farmers, food processors, brand owners and consumers to improve the sustainability as we go from "Farm to Fork".

Topics:
Energy (Sustainability & Environment)
Food (Sustainability & Environment)
Sustainable Supply Chains

The IMPACT of the Locally-Sourced FOOD Versus the MASS-Produced FOOD on Consumer Perceptions of Farm-to-Table and Carbon Footprint

- Conference: Food-Energy-Water Nexus
- Year: 2019
- Proceeding: Food-Energy-Water Nexus
- Group: Abstract Submissions
- Session:
  Lightning Round 2
- Time:
  Friday, December 6, 2019 - 2:20pm-2:25pm

Authors:
Li, Z., NYIT
Li, N., NYIT
Su, L., New York Institute of Technology
Wei, S., New York Institue of Technology
Sotiropoulos, V., New York Institute of Technology

The increased release of carbon is causing the climate to change negatively. These changes result in the melting of the glacier, the rising in sea levels and flooding. One industry that has tried to reduce its carbon footprint has been the food industry. The food industry is attempting to minimize their carbon footprint by reducing the distance between farms to table. Foods usually produced by local regional farms are commonly referred to as locally sourced foods. Initially, locally sourced food was expected to reduce the distance between farm to table and the carbon footprint in the production and distribution of the food process. In reality, the production and distribution of locally sourced food decreased the distance between farm-to-table but not the carbon footprint. Thus, research is needed to understand this discrepancy and reduce carbon footprint. This study rationalizes that consumers who become aware of the lack of impact in carbon footprint will pressure the locally sourced food industries to improve their sustainability practices that decrease carbon footprint. Therefore, this study evaluates consumer perceptions related to the farm-to-table and carbon footprint of locally sourced food versus mass-produced goods. Using a survey, 80 participants will be asked to assess their perceptions related to carbon footprint between mass-produced food and locally sourced food that is considered to be either high or low carbon footprint. The expectation is that consumer will associate locally sourced food with a shorter farm-to-table distance and a smaller carbon footprint. Results, discussion implications, limitations, and future research are discussed.

Topics:
Food (Sustainability & Environment)
Carbon Management

From National Indices to Regional Action-an Analysis of Food, Energy, Water Security in Ecuador, Bolivia, and Peru

- Conference: Food-Energy-Water Nexus
- Year: 2019
- Proceeding: Food-Energy-Water Nexus
- Group: Abstract Submissions
- Session:
  Lightning Round 2
- Time:
  Friday, December 6, 2019 - 2:25pm-2:30pm

Authors:
Mohammadpour, P., Pennsylvania State University
Grady, C., Pennsylvania State University

Food-Energy-Water (FEW) nexus frameworks may be well suited to provide tools for achieving the sustainable development goals by allowing for evaluation of multiple food, energy, and water priorities simultaneously. In terms of natural resources security, several quantification tools that consider accessibility and availability of resources also exist. Previous FEW security scholars have constructed quantification tools at national level, yet we hypothesize that these tools fail to articulate important regional variation within a country. In this research we tackle this challenge by developing methods to study the spatial patterns of FEW availability and accessibility at sub-national level in Ecuador, Peru, and Bolivia. The integrated approach in this study articulates how national statistics often misrepresent FEW security regional variation. Our tool also has the potential to be applied at multiple spatial scales allowing development actors to prioritize multiple natural resource development priorities worldwide.
Topics:
Food-Energy-Water Nexus

Treating Waste with Waste: Using Biochar to Recover Nutrients from AD Effluent

- Conference: Food-Energy-Water Nexus
- Year: 2019
- Proceeding: Food-Energy-Water Nexus
- Group: Abstract Submissions
- Session:
  Lightning Round 2
- Time:
  Friday, December 6, 2019 - 2:30pm-2:35pm

Authors:
Rodriguez Alberto, D., RIT
Trabold, T., RIT

The processing of food waste with animal manure through anaerobic digestion (AD) can be an effective waste management and energy recovery strategy. However, there is growing concern with the widespread deployment of AD, because managing its nutrient-rich effluent (digestate) has become an environmental challenge. Digestate is often separated into its solid and liquid fractions using mechanical processing such as a screw press. The solid fraction is often repurposed as cow bedding, while the liquid fraction is applied to crop fields as an alternative to synthetic fertilizers, which might lead to nutrient run-off and subsequent water contamination. We are developing a potential solution to this problem by processing the solid fraction of the effluent through thermochemical conversion (TC) to produce biochar, a porous carbonaceous material known for its adsorption capabilities. This material would be used to recover nutrients like nitrogen, phosphorus and potassium present in the liquid fraction of the effluent. The resulting nutrient-rich biochar can be applied as an alternative fertilizer that reduces potential water contamination. If an efficient integration of AD and TC processes is achieved, it can result in an enhancement of the system's economic viability and reduction of their associated environmental impacts.

Abstract:
P587865.pdf

Topics:
Biorefineries
Food (Sustainability & Environment)
Food-Energy-Water Nexus

To Meat or to Plant-Based Meat: The Impact of Consumer Resistance of Substitute Animal-Based Meat with Plant-Based Meat

- Conference: Food-Energy-Water Nexus
- Year: 2019
- Proceeding: Food-Energy-Water Nexus
- Group: Abstract Submissions
- Session:
  Lightning Round 2
- Time:
  Friday, December 6, 2019 - 2:35pm-2:40pm

Authors:
Wu, W., NYIT
Lu, X., NYIT
Meng, Z., New York Institute of Technology
Rodriguez, L., New York Institute of Technology
Sotiropoulos, V., New York Institute of Technology

Americans continue to consume large amounts of animal-based meat in their diet. Nonetheless, the traditional meat-based diet is no longer sustainable due to the health-related problems associated with consuming large quantities of animal based-meat. In fact, heart failure is the number one killer in America. The production of meat is also impacting the environment in an unsustainable fashion due to release carbon in the air. This has led consumers to develop negative perceptions of meat consumption and production. To help reduce the negative impact of producing and consuming animal-based meat on human health and the environment, some food companies have substituted it with plant-based meat. Plant-based meat has been shown to be a healthier alternative to animal-based meat and does not release the same level of carbon. Despite the positive impact of plant-based meat on human health and the environment, some consumers remain resistant to this substitution. Research is needed that will reshape consumers' apprehensive perception of a plant-based diet. Therefore, the goal here is to identify the reasons customers are reluctant to adopt plant-based meat. This study will focus on measuring nutrition illiteracy, level of carbon sustainability, carbon footprint and cost. The expectation in the present investigation, is that consumers who have less nutrition literacy, less knowledge and experience with the plant-based meat and who are not aware of the implication of unsustainable food production practices will more resistant to substitute animal-based meat with plant-based meat. The findings, implications, limitations and future areas of research will be also discussed.

Topics:
Sustainability & Environment

Impact of Electrifying Urban Food Delivery Trucks on the Environment and the Power Grid

- Conference: Food-Energy-Water Nexus
- Year: 2019
- Proceeding: Food-Energy-Water Nexus
- Group: Abstract Submissions
- Session:
  Lightning Round 2
- Time:
  Friday, December 6, 2019 - 2:40pm-2:45pm

Authors:
Elangovan, R., New York Institute of Technology
Jain, J., New York Institute of Technology
Esa, Y., City College of the City University of New York
Salmeron, L., City University of New York
Dong, Z., New York Institute of Technology
Mohamed, A., City College of the City University of New York

New York City's food distribution system is among the largest in the United States. Food is transported from six major distribution centers to various consumer point-of-sale locations by trucks. Freight is a large consumer of energy as well as an emitter of greenhouse gases. We present a case study on the impact of electrification of urban truck routes on the carbon emission and the power grid at the Gowanus district in Brooklyn New York. The study analyses the food consumption of the people in New York City based on their dietary pattern, and estimates the number and types of trucks needed to transport the food to this neighborhood. We compare the energy consumption, carbon footprint, and changes in electricity demand on the power grid for diesel versus electric trucks. Based on the predicted population growth, truck routes, point-of-sale locations, it is estimated that electric trucks may result in 40% less greenhouse gases than diesel trucks every year provided that electric trucks are charged using renewable energy resources. The impact on the power grid has been evaluated by the amount of energy needed for electric trucks per week and their peak power demand. The study shows that the current hosting capacity of the neighborhood under study is capable of handling food delivery electrification, without an immediate need for infrastructure upgrades.

Topics:
Food-Energy-Water Nexus

Food Waste to Energy: Leveraging Catalytic Hydrothermal Liquefaction

- Conference: Food-Energy-Water Nexus
- Year: 2019
- Proceeding: Food-Energy-Water Nexus
- Group: Abstract Submissions
- Session:
  Lightning Round 2
- Time:
  Friday, December 6, 2019 - 2:45pm-2:50pm

Authors:
LeClerc, H. O., Worcester Polytechnic Institute
Timko, M. T., Worcester Polytechnic Institute
Teixeira, A. R., Worcester Polytechnic Institute
Tompsett, G., Worcester Polytechnic Institute

Millions of tons of food waste are thrown into landfills annually.[1] Degradation products lead to a variety of detrimental wastes. Hydrothermal liquefaction (HTL) is an emerging technology in the waste-to-energy field that utilizes high temperature (T=300℃) and pressure (P=3000 psi) to convert this waste into a bio-oil. With appropriate in-situ or downstream upgrading, this oil is suitable for the transportation and heating industries. HTL's benefit is using complex, high water content feedstocks without inefficient drying steps.

Homogeneous catalysts have been utilized with HTL, however, pose process complexity and expense with separation and neutralization. Recently, solid catalysts, with acid- base characteristics have been used successfully with in-situ upgrading of food waste to bio-oil. [2,3]. Hydroxyapatite is a crystalline mineral found naturally in bones and makes an attractive catalyst due to its tunable acid-base sites and potentially renewable sources. This relatively inexpensive catalyst has preliminarily shown to result in the highest oil yield from food waste. In this work, a near supercritical liquid phase reactor is used with a food waste/catalyst slurry to drive the complex reactions. A bio-oil is generated with higher heating values approaching the range of diesel fuel and the catalyst shows excellent hydrothermal stability and regeneration. Additionally, discussion of producing bio-oil in a flow-based system will be discussed.

Topics:
Food (Sustainability & Environment)
Catalysis
Biofuels (Energy)

Invited Talk: Designing Multifunctional Processes to Direct the Synthesis of Clean Energy Carriers with Inherent CO2 Capture and Conversion from Heterogeneous Carbonaceous Feedstocks

- Conference: Food-Energy-Water Nexus
- Year: 2019
- Proceeding: Food-Energy-Water Nexus
- Group: Abstract Submissions
- Session:
  Session 3: Biogas and Infrastructure
- Time:
  Friday, December 6, 2019 - 3:10pm-3:30pm

Authors:
Gadikota, G., Cornell University
Yin, T., Cornell University
Gao, X., Cornell University

Rising need to treat carbonaceous food waste in the urban environment calls for advancing the science of multifunctional processes to direct the synthesis of clean energy carriers such as H_2 with inherent carbon removal. One such process involves combining gasification of these carbonaceous feedstocks with the water gas shift reaction to produce a clean energy carrier such as H_2 and CO_2. The reactive separation of CO_2 from H_2 is achieved by using earth abundant sorbents being calcium or magnesium to produce water insoluble calcium or magnesium carbonate, also known as carbon mineralization. Integrating these reactions directly with the water gas shift reaction has been shown to be thermodynamically favorable. Further, the presence of water kinetically favors these reactions. In this study, we discuss pathways to aid the reactive separation of CO_2 and direct the synthesis of H_2.

Topics:
Energy (Sustainability & Environment)
Environmental Sustainability
Water (Sustainability & Environment)

Anaerobic Wastewater and Environmental Microbial Communities Readily Demethylate Methoxy Aromatic PPCPs

- Conference: Food-Energy-Water Nexus
- Year: 2019
- Proceeding: Food-Energy-Water Nexus
- Group: Abstract Submissions
- Session:
  Session 3: Biogas and Infrastructure
- Time:
  Friday, December 6, 2019 - 3:30pm-3:47pm

Authors:
Wolfson, S. J., Albert Einstein College of Medicine
Porter, A. W., Rutgers University
Young, L. Y., Rutgers University

Pharmaceuticals and personal care products (PPCPs) can be metabolized by microbes during wastewater treatment, releasing both intact PPCPs and their metabolites to the environment. There is an urgent need to understand PPCP transformation however the multitude of chemicals makes studying each microbial metabolism an arduous task. We propose that predicting microbial PPCP metabolism during wastewater treatment is best understood in the context of shared chemical structure. We recently described a methanogenic enrichment culture from anaerobic wastewater sludge that readily demethylates the common pharmaceutical naproxen to 6�-O-�desmethylnaproxen, with no loss of the metabolite observed. In this culture, methane was generated at stoichiometric amounts to naproxen demethylation, indicating that the methyl group removed from naproxen stimulates a complex microbial community. The natural methoxy aromatic compounds syringic and vanillic acids were also readily demethylated, and this shared metabolism prompted an examination of other methoxy aromatic xenobiotic compounds. In the present study we challenged the methanogenic wastewater culture with the methoxy aromatic UV filter oxybenzone and observed demethylation to benzophenone�1, which itself is an estrogenic contaminant. A second culture from sulfidic marine sediment was established to simulate anoxic environmental sediment receiving pharmaceutical contamination. Again, the natural and PPCP methoxy aromatic compounds were readily demethylated with accumulation of the metabolites. This study provides experimental support to a systematic understanding of methoxy aromatic PPCP transformation, those investigated here and those not yet investigated. It suggests that methoxy aromatic PPCPs are transformed to demethylated metabolites during anaerobic wastewater treatment and the environment, and their metabolites may accumulate.

Topics:
Water (Sustainability & Environment)
Understand Hazard & Risk
Biochemical Processes

Transforming Rural Community Economy from Linear to Circular One Using Biogas Production Technology

- Conference: Food-Energy-Water Nexus
- Year: 2019
- Proceeding: Food-Energy-Water Nexus
- Group: Abstract Submissions
- Session:
  Session 3: Biogas and Infrastructure
- Time:
  Friday, December 6, 2019 - 3:47pm-4:04pm

Authors:
Momodu, A. S., Obafemi Awolowo University
Adepoju, T. D., Obafemi Awolowo University

Twin sources - fossil fuel usage and poor waste management – that contribute to greenhouse gas emissions can be combated using the concept of circular economy through using biogas technology. This approach will allow waste conversion to meet rural energy demand and also reduce dependence on fossil fuel use. This study presents technical and business model approaches to upscale a laboratory experimental procedure for biogas production through anaerobic digestion using vegetal wastes at mesophilic temperature of $35^0C – 40^0C$. Inoculated substrate is up-scaled using molarity concept. Waste required is 13.3 tons for $67m^3$ digester for 30-day hydraulic retention time to produce 39.13 m^3, with daily organic loading rate is 0.44 ton of total solids. Avoided CO_2 emissions from liquefied petroleum gas, kerosene, diesel, fuel wood is 9.69kg, 37.49kg, 23.147kg, 181.65kg respectively. Total investment cost is estimated at $83,123. Life cycle cost (LCC) and return streams are estimated based on three scenarios with a 20-year life-cycle. LCC for digestate in Scenarios A and B is the same at $0.13/kg and Scenario C at $0.09/kg. For electricity generation, LCC is $0.008/kBTU and $0.01/kBTU in Scenario B and C respectively, cooking is $0.14/kg and $0.19/kg in Scenario A and C respectively. Return streams for each of the three scenarios are estimated at $116,373.36. User-centered design is used to develop the business model. The project fits well as a means to transform rural community economy from linear to circular system through producing clean and renewable energy, managing natural resources and providing jobs.
Topics:
Alternative Energy
Environmental Sustainability
Energy (Sustainability & Environment)

An Analytical Framework for Assessing Biogas Potential from Consumer Food Waste in US Cities

- Conference: Food-Energy-Water Nexus
- Year: 2019
- Proceeding: Food-Energy-Water Nexus
- Group: Abstract Submissions
- Session:
  Session 3: Biogas and Infrastructure
- Time:
  Friday, December 6, 2019 - 4:04pm-4:21pm

Authors:
Beagle, E. A., UT Austin
Glazer, Y. R., UT Austin
Webber, M. E., UT Austin

Consumer-side food waste—cited anywhere from 20% to 40% of all food production—is a potential renewable fuel source that could offset significant energy needs in urban areas. Anaerobic digestion is a proven method by which biogas can be produced from food scraps. Currently, logistical challenges and costs make implementing city-wide consumer food waste collection and use limited.

However, many US cities already have composting programs which collect food waste and compost it to make fertilizer and high quality soil. In addition, many cities have ambitious goals and initiatives to significantly reduce their waste and greenhouse gas emissions in the upcoming decades. The use of food waste to produce biogas would help cities to simultaneously meet both of these objectives. In this study we establish an analytical framework for assessing the energy availability from consumer food waste in US cities and examine the potential for displacing existing natural gas usage with biogas derived from consumer food waste through anaerobic digestion. Results will include an estimation of available consumer food waste, potential to meet demand using produced biogas via a comparison with existing natural gas usage, and potential emission reductions from both reducing food waste in landfills and displacing consumer natural gas usage. Results will help inform city policies for establishing various food waste collection and use programs.

Topics:
Food-Energy-Water Nexus
Energy (Sustainability & Environment)
Biofuels (Energy)

Infrastructure Expansion Planning Under Food-Energy-Water Nexus Considerations

- Conference: Food-Energy-Water Nexus
- Year: 2019
- Proceeding: Food-Energy-Water Nexus
- Group: Abstract Submissions
- Session:
  Session 3: Biogas and Infrastructure
- Time:
  Friday, December 6, 2019 - 4:21pm-4:38pm

Authors:
Avraamidou, S., Imperial College London
Allen, R. C., Texas A&M Energy Institute, Texas A&M University
Pistikopoulos, E. N., Texas A&M Energy Institute, Texas A&M University
Daher, B., Texas A&M University
Di Martino, M., Texas A&M University
Cook, J., Texas A&M University

Due to global population growth, economic development, and urbanization, the demands for food, energy, and fresh water are rapidly increasing, exerting pressures on existing supply chains and urban infrastructure. These pressures are exacerbated by climate change impacts and delayed policy response. As these demands grow, the stresses and interdependences between these resources become more apparent. A holistic approach for the optimal decision-making for energy and water infrastructure planning that build on a Food-Energy-Water Nexus (FEW-N) framework is essential to meet the increasing utility demands. Such a decision-making approach is complex and involves challenges related to: (i) involvement of different stakeholders with often conflicting objectives, (ii) the intrinsic intermittent nature of renewable energy sources; and (iii) stochastic energy and water demand profiles.

In this work, we present a novel FEW-N superstructure-based representation and multi-objective optimization framework for energy and water infrastructure planning. The FEW-N problem is posed as a two-stage stochastic mixed-integer linear program that minimizes the capital expenditures, operational cost, water usage and land competition of the FEW-N system. Planning decisions taken into consideration include the ability to construct new power plants, energy storage units and water treatment plants, while scheduling decisions involve the operating level of the plants, how the energy is allocated within the system, and when energy is stored and released from storage devices. We illustrate the effectiveness of the infrastructure expansion framework through the use of a case study that spans the ERCOT region of Texas.

Topics:
Energy (Sustainability & Environment)
Water (Sustainability & Environment)
Environmental Sustainability

Transforming Wastewater Systems As Community Resource

- Conference: Food-Energy-Water Nexus
- Year: 2019
- Proceeding: Food-Energy-Water Nexus
- Group: Abstract Submissions
- Session:
  Session 3: Biogas and Infrastructure
- Time:
  Friday, December 6, 2019 - 4:38pm-4:55pm

Authors:
Daw, J., NREL
Stout, S., NREL

Wastewater systems exemplify the intersection of energy, water, land-use, and food production and are increasingly being viewed for the resources they provide. These systems are unique in that they possess opportunities to reduce community energy use and generate on-site energy as a byproduct of their operations. Further, biosolids generated as a result of the wastewater process have beneficial uses as land amendments for agriculture and landscaping purposes, providing additional value streams that can further reduce utility operating costs.

Energy use can account for as much as 10% of a local government's annual operating budget. While communities closely track costs associated with treating wastewater, more granular information on the specific energy, transportation, and land use implications associated with these costs is less understood. There is also poor availability of data to assess the effect of policy and operational decisions related to water conservation and reuse, treatment process optimization, new technologies, on-site energy generation, transportation, and land application of biosolids.

In this work, NREL brought together a diverse energy dataset from organizations such as the Water Environment Federation, Water Research Foundation, Electric Power Research Institute, NYSERDA, EPA, and universities to understand the broader implications and opportunities related to wastewater systems. This study considered the impact of domestic trends such as population growth, urbanization, new industries, aging infrastructure and climate change on wastewater systems and their broader implications on energy, water, land and food production. The resulting tool that was developed can be used to help inform policy and operational decisions for communities

Topics:
Alternative Energy
Energy (Sustainability & Environment)
Food-Energy-Water Nexus

Employing Decision Analysis to Evaluate Nontraditional Management Pathways for Produced Water from Oil and Gas Extraction

- Conference: Food-Energy-Water Nexus
- Year: 2019
- Proceeding: Food-Energy-Water Nexus
- Group: Abstract Submissions
- Session:
  Session 3: Biogas and Infrastructure
- Time:
  Friday, December 6, 2019 - 4:55pm-5:12pm

Authors:
Glazer, Y. R., UT Austin
Webber, M. E., UT Austin
Beagle, E. A., UT Austin

Produced water is a common byproduct of oil and gas (O&G) production that must be properly managed to avoid damage to the environment and human health. Currently, the majority of produced water is disposed underground via deep well injection. In addition, O&G development can be limited when there is insufficient disposal capacity or wastewater management strategies. Over time, treatment and reuse of the produced water for beneficial uses has become more commonplace yet the majority of produced water is still disposed downhole.

In this study, we built a framework employing decision analysis methodology to determine the economic tradeoffs of several produced water management pathways. Nontraditional pathways including piping and selling treated water to a municipality, irrigating terrestrial energy crops onsite, irrigating greenhouse crops, among others, were compared to more typical management strategies of discharging to a nearby stream or underground injection. To assess the water management pathways, the model incorporates key parameters including CAPEX, OPEX, potential revenue, and uncertainty for each option. The methodology is illustrated using data found in literature along with field data and conditions from an oil and gas site in central Wyoming. We determined that the management pathway with the highest expected value is discharging to a nearby stream followed closely by irrigating an onsite energy crop. This result is primarily due to the lower treatment and brine disposal costs associated with these two options. This work also determined the commodity prices required for each of the management pathways to achieve profitability.

Topics:
Energy
Water (Sustainability & Environment)
Food-Energy-Water Nexus

"Chromash": Adsorping Challenges, Desorping Sustainability

- Conference: Food-Energy-Water Nexus
- Year: 2019
- Proceeding: Food-Energy-Water Nexus
- Group: Poster Submissions
- Session:
  Poster Session
- Time:
  Thursday, December 5, 2019 - 5:00pm-6:00pm

Authors:
Omar, M., University Of Khartoum
Osman, B., University Of Khartoum
Osama, A., University Of Khartoum
Awad, A. S., University Of Khartoum

"ChromAsh" is an invented powder product from treated petroleum coke bottom ash that has been scientifically proven to adsorb waste contaminants from tannery effluents with 99.99% effectiveness under variable process conditions. The process is non-complicated including only, filtration, adsorption and desorption columns. Extremely reduced fixed and operational cost when compared with current technologies. The Process offers opportunity to recycle 100% of tanning dye, tanning water and sure the adsorbent itself. In last September, the new technology is approved by Sudanese Standards Metro-logy Organization (SSMO).

"ChromAsh" is expected to be tomorrow's solution for continuous tannery discharges harming both humans and environment, specially when regarding the traditional tanning process which is the dominant in 78%of African and Asian countries.**"ChromAsh"** with countless economic, environmental and health-protection advantages to both producers and consumers in the energy and tanning industries is searching for a sponsor for more scientific development and large-scale implementation.

Topics:
Sustainability & Environment
Sustainable Engineering
Specialty Chemicals

Alternative Approach to the Design of Toilets for Optimal Water Consumption

- Conference: Food-Energy-Water Nexus
- Year: 2019
- Proceeding: Food-Energy-Water Nexus
- Group: Poster Submissions
- Session:
  Poster Session
- Time:
  Thursday, December 5, 2019 - 5:00pm-6:00pm

Authors:
Peñaranda Palacios, C. N., EAN. university
Pinzón Bustamante, L. D., EAN University

Recent studies and reports by the United Nations Program (UN) according to the environment, reports the water resource is one of the most critical resources. Colombia is made up of 238 municipalities, where 28% of the national territory has water rationing. For example, different countries look for alternatives to work on the improvement of systems for urban prevention and planning in order to create sustainable cities with water quality. The capital of Colombia, Bogotá, has losses of up to 42% of the water according to the studies of the Department of National Planning. Because of this, relevant alternatives are required to achieve the city's water consumption and thus, move towards a smart city model. In Colombian households it is estimated that the average consumption of a household is 10.76 cubic meters of water, strata 4,5 and 6, the areas where there is a greater consumption are in bathrooms and showers, where it is estimated that 40% of household drinking water is spent on sanitary discharges. Due to this, ConIntA research seedbed has worked on the models of affected toilets to meet the needs of the population and the relevant characteristics for the design and invention of a new design of the toilets, where it is proposed to make use of the fundamental physical principle, the vacuum, to handle the discharges together with pressure systems.

Topics:
Water (Sustainability & Environment)
Regulations
Design/Green Engineering

An Analysis of the Possible Financial Savings of a Carbon Capture Process through Carbon Dioxide Absorption and Geological Dumping

- Conference: Food-Energy-Water Nexus
- Year: 2019
- Proceeding: Food-Energy-Water Nexus
- Group: Poster Submissions
- Session:
  Poster Session
- Time:
  Thursday, December 5, 2019 - 5:00pm-6:00pm

Authors:
Ssebadduka, R., Kyushu University

Herein, we discuss possible ways to reduce the cost of carbon dioxide (CO_2) sequestration with a special focus on the process and the solvent used. Modifications to the process to eliminate the stripper section and focus on just the CO_2 adsorbing and geological dumping (CAGD)can lead to significant reductions in the sequestration cost per tonne of CO_2 compared with ordinary CO_2 capture and geological storage (CCS) processes. In the case of CAGD, savings of steam used in the ordinary CO_2 capture process can go up to US$12.7per ton of CO_2 captured and additional savings on the waste disposal cost of US$175/tonne of waste can be made. More savings on the energy costs for compression and cooling of the captured CO_2 gas can be realized if the absorbent and flue gas/CO_2 slurry is directly dumped in a geological formation. A change of the capture solvent can also make this process better economically by using the proposed substitute. Many commercially available alternatives to monoethanolamine (MEA) have been presented in this report by mainly focusing on how their loading capacity and cost compare. Aqueous Sodium carbonate (Na_2CO_3) has been proposed as the best material for use in the CAGD CO_2 absorbing process based on the economic advantages it presents.

Topics:
Carbon Capture & Storage

Carbon Capture & Storage

Carbon Capture Using Industrial Hydroxides for EOR-Opportunities,Challenges and the Way Ahead

- Conference: Food-Energy-Water Nexus
- Year: 2019
- Proceeding: Food-Energy-Water Nexus
- Group: Poster Submissions
- Session:
  Poster Session
- Time:
  Thursday, December 5, 2019 - 5:00pm-6:00pm

Authors:
Shreyash, N., Rajiv Gandhi Institute of Petroleum Technology

To alleviate CO_2 emissions, the most dominating factor towards increasing global warming, carbon capture and storage(CCS) proves to be promising. Combining various CCS technologies, the use of cheaper and abundant fossil fuels still elevate the industrial growth. In the meantime, CO_2 emissions also witness a decrement. Out of various amine-based solvents, mono-ethanol-amine(MEA) is the only one in use in various post-combustion plants for carbon capture from flue gases. The regular refilling and repacking of MEA to make up for losses as a result of high volatility and solvent degradation is one limitation of MEA.
Hydroxides like sodium hydroxide can rectify the existing drawbacks because NaOH is known to have higher CO_2 absorption than MEA. But NaOH has its deficiencies like it cannot be regenerated post-capture and $NaHCO_3$ is the reason for it as at 160-degree-Celcius it decomposes to give Na_2CO_3, H_2O and CO_2. At very severe conditions, Na_2CO_3 forms Na_2O. The severity of these conditions is as high as providing 800-degree-Celcius temperature.
Usage of NaOH is preferred for flooding and oil recovery in reservoirs after the capture because it doesn't require recovery. NaOH provides suitable milieu for micelle formation when used in alkali-surfactant-polymer. By injecting CO_2 laden NaOH, CO_2 sequestration is eased and extra oil is produced due to alteration of oil miscibility because of the presence of bubbled CO_2. Further substitution of NaOH with other hydroxides is also a possibility which can be examined based on oil-field conditions. Exploring these possibilities may enhance CO_2 capture and oil recovery.
Topics:
Carbon Capture & Storage
Hydrogen Production & Storage

Carbon Dioxide Solubilities in Decanoic Acid-Based Hydrophobicdeep Eutectic Solvents

- Conference: Food-Energy-Water Nexus
- Year: 2019
- Proceeding: Food-Energy-Water Nexus
- Group: Poster Submissions
- Session:
  Poster Session
- Time:
  Thursday, December 5, 2019 - 5:00pm-6:00pm

Authors:
Gupta, V., Rajiv Gandhi Institute of Petroleum Technology

To alleviate drawbacks like high toxicity, high cost, non biodegradability etc, deep eutectic solvents (DES's) were found and are the modern solvents that have the caliber to ameliorate these fallibilities and we use these solvents to examine the potential of CO2 as a capturing agent. The preparation of DES takes mixing of decanoic acid and quaternary ammonium salt with variegated halide anions and alkyl chains.Varying temperature from 298 to 323 K and at a maximum pressure of 2MPa, six unique hydrophobic DES's were put in to be discerned and outcomes have been innumerated after solubility data with existent hydrophilic DES's, contemporary physical solvents and fluorinated ionic liquids have been tallied. All this has been done with a motivation to wander into a new arena of measuring CO2's solubility in DES's and such experiments prove the potency of CO2 as a capturing agent. The measured CO2 solubilities have nuance differences with those of fluorinated ILs but CO2 absorption lies in the bracket of those of nonpolar solvents.

Topics:
Carbon Capture & Storage
Carbon Management
Biochemical Processes

Chinese Food in China Vs. Chinese Food in America: Chinese and American Perceptions of Healthiness

- Conference: Food-Energy-Water Nexus
- Year: 2019
- Proceeding: Food-Energy-Water Nexus
- Group: Poster Submissions
- Session:
  Poster Session
- Time:
  Thursday, December 5, 2019 - 5:00pm-6:00pm

Authors:
Cai, Y., New York Institute of Technology
Sotiropoulos, V., New York Institute of Technology

ABSTRACT

American has an obesity problem. As a result, they often look for healthier food options. One such option is Chinese food. Some research suggests that Americans may erroneously think Chinese food is healthier than it may actually be. They equate the lack of obesity in Chinese as a

product of their diet. This can lead Americans to adopt a Chinese diet as a healthier alternative. But in fact, the Chinese food made in America may not be as healthy as it is perceived. The presence of saturated fats and sugar can make such food unhealthy. The present study investigates this assumption using a survey approach of 50 participants (i.e.,25 Americans, 25 Chineses) in the United States. The expectation in this study is to show that most Americans view Chinese food made in American as healthier than

it actually is and also, to show that Chinese view Chinese food made in

America as less healthier than Americans views it. The results, findings, implications, limitations, and future research will be also be discussed

Topics:
Food (Sustainability & Environment)
Business Strategy and Economics (Sustainability & Environment)

Do Americans Have a Misguided Interpretation of American Raised Food

- Conference: Food-Energy-Water Nexus
- Year: 2019
- Proceeding: Food-Energy-Water Nexus
- Group: Poster Submissions
- Session:
  Poster Session
- Time:
  Thursday, December 5, 2019 - 5:00pm-6:00pm

Authors:
Mkhonto, N., NYIT
Chery, A., NYIT
Fernandez, S., NYIT
Sotiropoulus, V., NYIT

ABSTRACT

The number of food deserts in America has increased. Food deserts are defined as parts of the country vapid of fresh fruit, vegetables, and other healthful whole foods, usually found in impoverished areas such as low-income neighborhoods. Areas that are heavily composed with non-nutritional food options like fast food and highly processed food. (paraphrase) The lack of nutritional food options in these areas can have negative health consequences such as obesity, heart problems, diabetes to name a few. Nonetheless, it is unclear if the people that reside in such areas are aware just how nutritionally deficient their environment is. Furthermore, it is unclear if these areas are so nutritionally deficient that they rival or even worse than areas in of more impoverished nation. Therefore, the goal of this study is to investigate food nutritional deficiency between Americans who live in food deserts with people in a developing nation (i.e., South Africa). In the present study, we expect to find that individuals who reside in food deserts have food options that are so nutritional deficient that even people who reside in less developed nations have more nutritional options. Using a qualitative interview approach 15 participants will be interviewed to determine if there is evidence to support this claim. The results, findings, implications, limitations, and future research will also be discussed.

Topics:
Food (Chemicals & Materials)

Food Waste Separation & Depackaging:Capturing Organic Residuals for Nutrient Repurposing, Compost and/or Renewable Energy

- Conference: Food-Energy-Water Nexus
- Year: 2019
- Proceeding: Food-Energy-Water Nexus
- Group: Poster Submissions
- Session:
  Poster Session
- Time:
  Thursday, December 5, 2019 - 5:00pm-6:00pm

Authors:
Rossen, C., Ecoverse

Food waste separation of organic material from its packaged containment is no longer being overlooked, but rather becoming a necessity for companies, municipalities and other organizations. Food (and other organic) waste depackaging leads to the maximization of animal nutrition, composting and anaerobic digestion while minimizing dependency on landfills, therefore satisfying regulatory compliance. Specifically, depackaging systems are seeking to make a significant contribution to industries by:

• Increasing the capacity for animal nutrition, composting and anaerobic digestion of source segregated, prepackaged food and organic waste
• Encouraging the demand for quality food and organic waste outputs in the marketplace
• Identifying and opening new markets for animal nutrition, composting and anaerobic digestion outputs
• Recycling organic waste streams for continued repurposing
• Reduce environmental impact of discarded organic waste

Separation technology and equipment have made it possible to remove organic contents from its packaging for recycling/repurposing. This separation of materials leads to the increased utilization of animal nutrition, composting and anaerobic digestion. My presentation will explore the ability to separate organics from its packaging onsite at food manufacturing, retail, municipalities, stand alone facilities and transfer stations as a means to divert organic waste from landfills and repurpose/recycle the material through animal nutrition, composting and anaerobic digestion.

Topics:
Food (Sustainability & Environment)

GIS-Based Assessment of Regional Biomass Potentials for Heat and Power Generation at the Example of Ludwigsburg County, Germany

- Conference: Food-Energy-Water Nexus
- Year: 2019
- Proceeding: Food-Energy-Water Nexus
- Group: Poster Submissions
- Session:
  Poster Session
- Time:
  Thursday, December 5, 2019 - 5:00pm-6:00pm

Authors:
Padsala, R., University of Applied Sciences - HfT Stuttgart
Bao, K., University of Applied Sciences - HfT Stuttgart
Kesnar, C., Univeristy of Applied Sciences - HfT Stuttgart
Coors, V., Univeristy of Applied Sciences - HfT Stuttgart
Schr?ter, B., Univeristy of Applied Sciences - HfT Stuttgart

The county of Ludwigsburg is located in the centre of Baden-Württemberg, in Germany's south-western corner. It covers an area of approximately 265 square miles, with a population of about 550,000 inhabitants, spread across 39 municipalities. In 2013, the county set itself a goal of reducing greenhouse gas emissions by 90 percent until 2050 through inceased energy efficiency and increasing the share renewable energy sources for heat and power generation.
In light of this, an assessment of the biomass potential for heat and power generation for the county of Ludwigsburg is presented here. Our study uses Digital Landscape Models (DLM) in the open CityGML data format to calculate the local biomass potential, and couples these with our workflow-based urban simulation platform SimStadt. Since the landscape type for each relevant polygon in the DLM is known (wood, shrub, agriculture), crops can be allocated to the agricultural polygons based on statistical data. In a next step, biomass potentials per polygon are catogorized into for example energy wood, bioethanol, biogas, or vegetable oil.
In Ludwigsburg, the local biomass potential amounts to about 761 GWh/a, or 4% of total energy demand (based on 2012 demand data). This compares favourably to an actual biomass share of 2% of total energy demand (in 2012).

Topics:
Energy (Sustainability & Environment)
Sustainability & Environment
Research and Development

Meal-Kit Delivery: The Trade-Offs of Food Waste and Energy Consumption

- Conference: Food-Energy-Water Nexus
- Year: 2019
- Proceeding: Food-Energy-Water Nexus
- Group: Poster Submissions
- Session:
  Poster Session
- Time:
  Thursday, December 5, 2019 - 5:00pm-6:00pm

Authors:
Davidson, T., United States Military Academy
Webber, M., University of Texas at Austin
Gee, I., University of Texas at Austin
Speetles, B., University of Texas at Austin

This work shares a model that was developed to compare the energy requirements of meal-kit delivery systems to conventional grocery shopping. Meal-kit services can reduce food waste because the kits pre-portion ingredients for each recipe, thereby saving energy. However, the supply chain and packaging requirements of meal-kit delivery are different than those for grocery stores, potentially offsetting any energetic benefits of reduced food waste. If meal-kit delivery replaces some trips to the grocery store, then transportation-related savings might be significant. The tradeoffs of these competing effects are non-obvious, so mass and energy balances were used to assess embedded energy in both pathways. The model was illustrated under representative operating conditions for a consumer in Austin, Texas using Monte Carlo simulation. Both per-meal and per-week, a meal-kit delivery service meal is more energy intensive than procuring the same meal from conventional grocery stores primarily due to single-use packaging. Consumer transportation to the grocery store was also found to be particularly energy intensive. These results suggest that the energetic requirements of meal-kit delivery services could be reduced such that they are less than conventional grocery shopping if reusable or low-impact packaging is used, and if the delivery services are able to reduce the number of weekly trips to the grocery store. This work was published in the Journal of Cleaner Production, Volume 236, 1 November 2019, https://doi.org/10.1016/j.jclepro.2019.07.062.

Topics:
Food-Energy-Water Nexus
Food (Sustainability & Environment)
Energy

Mitigating Polymer Absorption in Porous Silica Formations with Surfactants & Silicananofluids: A Viscosity, Temperature and Rheology Study Investigation

- Conference: Food-Energy-Water Nexus
- Year: 2019
- Proceeding: Food-Energy-Water Nexus
- Group: Poster Submissions
- Session:
  Poster Session
- Time:
  Thursday, December 5, 2019 - 5:00pm-6:00pm

Authors:
Tiwary, S. K., Rajiv Gandhi Institute of Petroleum Technology

While the use of nanotechnology has been has been explored in many applications and of late, the application of nanotechnology in the various upstream petroleum activities like exploration, drilling, production and distribution. It has been observed that on the addition of nanoparticles to the fluids alter the properties of the fluid causing wettability alternation and impacting advanced drag reduction, strengthening sand consolidation, reducing the interfacial tension and increasing the mobility of the capillary-trapped oil. The focus of our current study has been the use of silica nanoparticles to mitigate the absorption of polymer in a porous silica formation. Conventionally used polymers to increase the viscosity the injection water are PAM, HPAM and Xanthan Gum. The use of polymer in an injection fluid causes the increment of oil from a formation. However, the polymer has been found to adsorb onto reservoir rocks, causes precipitation, and results in the changes to the rheological properties of the fluid. Hence, we have explored the use of nanoparticles in mitigating the adsorption of water-soluble polymers on the solid surfaces of sandstone has been investigated. Our study has focused on the role of polymer and its resulting absorption in a porous media on the viscosity, temperature and salinity of solutions of polyacrylamide polymers with different nanoparticle & surfactant concentrations. Our studies have shown that the use of polymer solutions containing silica nanoparticles have significantly less adsorption based on weight percent than conventional samples. This attribute is hence desirable for surface absorption.

Keywords: Polymer Absorption, Enhanced Oil Recovery, Nanoparticles, Surfactant, Clays, Sandstones

Topics:
Nanomaterials
Absorption

Optimal Allocation of Renewable Energy Generation and Storage Systems

- Conference: Food-Energy-Water Nexus
- Year: 2019
- Proceeding: Food-Energy-Water Nexus
- Group: Poster Submissions
- Session:
  Poster Session
- Time:
  Thursday, December 5, 2019 - 5:00pm-6:00pm

Authors:
Cook, J., Texas A&M University
Avraamidou, S., Imperial College London
Allen, R. C., Texas A&M Energy Institute, Texas A&M University
Pistikopoulos, E. N., Texas A&M Energy Institute, Texas A&M University

There is an increasing demand for renewable power capacity in Texas due to the growing population statewide. However, increasing the amount of power supplied to the grid through renewable technologies can place stresses on the food, energy and water resources in the area. The aim of this research project is to model and optimize the allocation of renewable energy generation and storage technologies considering various objectives, such as cost and environmental impact, and scenarios, such as varying energy demands and land and water use constraints. Detailed high fidelity modeling of various energy generating processes, including solar farms, wind farms, and biomass, was performed. Nonlinear optimization was then used to obtain the energy output profiles of the various renewable energy generation technologies. The derived profiles were then used as inputs to a linear optimization program to obtain the optimal allocation for each scenario through mixed-integer linear optimization. The results of this model indicate that varying the restrictions placed on the model, such as limiting land area or water availability, can greatly affect the optimum energy production and storage methods. The proposed methodology has been adopted for a water-stressed region in south-central Texas. The outcomes of this research stress the significance of geographic factors, the energy demand profiles and resource availability, such as water and land, on the transition towards renewable energy generation.
Topics:
Alternative Energy
Water (Sustainability & Environment)

Soft Computing Tool for Aiding the Integration of Renewable Energy Systems in Isolated Areas of Colombia

- Conference: Food-Energy-Water Nexus
- Year: 2019
- Proceeding: Food-Energy-Water Nexus
- Group: Poster Submissions
- Session:
  Poster Session
- Time:
  Thursday, December 5, 2019 - 5:00pm-6:00pm

Authors:
Rosso-Ceron, A. M., UNIVERSIDAD INDUSTRIAL DE SANTANDER
Kafarov, V., Industrial University of Santander

In Colombia, several entities have studied the possibilities of renewable energy development in isolated areas. These studies suggest that the decision of choosing sustainable alternatives of energy must include, resources, as well as, sustainability dimensions. Thus, it is necessary to have planning tools that offer to manage energy system simulation and optimization models. Although, these tools are not enough to make the decision, because they do not consider other factors that affect the decision of choosing a sustainable system.

To address these problems, there are systems known as soft computing, among them, the fuzzy modelling makes decision making more flexible, and reliable. Besides, the development of computer facilities for the energy system allows a drastic reduction in the models' CPU time, making room for a continuous improvement in modelling and solving strategies.

Thus, this study contributes to the developed of a soft computing tool for the integration of renewable energy systems. The tool consists of a forecasting sub-tool for assessing quantitative and qualitative data. These data are input for a fuzzy multi-objective decision model developed in GAMS, where the total present value and the CO_2 emissions of the system are minimized for obtaining alternatives of Pareto. These alternatives are used as inputs for a fuzzy multi-attribute model developed in MATLAB where the CO_2 emissions and the total present value, with social, technologic, and environmental criteria are used as inputs for choosing the most sustainable alternative in a tested case of Chocó.

Topics:
Energy (Sustainability & Environment)
Energy Production
Sustainable Supply Chains

Understanding and Conceptualizing How Urban Green Blue Infrastructure Affects the Food, Water, and Energy Nexus: A Synthesis of the Literature

- Conference: Food-Energy-Water Nexus
- Year: 2019
- Proceeding: Food-Energy-Water Nexus
- Group: Poster Submissions
- Session:
  Poster Session
- Time:
  Thursday, December 5, 2019 - 5:00pm-6:00pm

Authors:
Meng, F., Yale University

The interactive dynamics in the food, water, and energy system as a nexus (FWEN) are critical to the sustainable development of global cities, which can be mediated by green and blue infrastructure (GBI) in the urban area. Here we provide a comprehensive literature review to examine how GBI affects FWEN in urban centers which is lacking current literature. In order to do this, we undertake a systematic review of the literature using a meta-analytic approach and topic modelling. Based on our synthesis, we develop a conceptual framework of the key links between urban GBI and FWEN and the direction and magnitude of the relationship. We found that GBIs can have positive and negative effects on urban FWEN. The negative effects can and should be addressed by policy making. In cities with different size/income/in different regions, GBIs can have heterogeneous effect on FWEN and should be adapted to specific localized polity targets. Finally, current analysis focuses on isolated analysis, lacking integrated analysis in develop decision making supporting tool which should be addressed in the future research.
Topics:
Food-Energy-Water Nexus
Sustainability & Environment

Utilizing Novel Technologies for Extraction of Bioactive Compounds from Mandarine Juice Production Waste

- Conference: Food-Energy-Water Nexus
- Year: 2019
- Proceeding: Food-Energy-Water Nexus
- Group: Poster Submissions
- Session:
  Poster Session
- Time:
  Thursday, December 5, 2019 - 5:00pm-6:00pm

Authors:
Karlovi?, S., Faculty of Food Technology and Biotechnology
Bosiljkov, T., Faculty of Food Technology and Biotechnology
Dujmi?, F., Faculty of Food Technology and Biotechnology
Brn?i?, M., Faculty of Food Technology and Biotechnology
Marelja, M., Faculty of Food Technology and Biotechnology
?kegro, M., Faculty of Food Technology and Biotechnology
Je?ek, D., Faculty of Food Technology and Biotechnology

Mandarine peels are waste from fruit juice production and, as such, have a low price and high potential to be further processed using novel food processing technologies such as high hydrostatic pressure or high – intensity ultrasonics. Environmentally friendly and low-cost raw materials and technologies are forcing the industry to develop new methods to guarantee the sustainability of the food chain. From this perspective, food waste made in the fruit juice industry is of great interest since the research shows that the peel contains a high concentration of bioactive compounds. Those compounds could be extracted and further used in food or pharma industry. The primary objective of this research was extraction with supercritical carbon dioxide and comparison of obtained results with other extraction techniques such as conventional extraction and extraction with high-intensity ultrasound.

Results show that the ultrasound-assisted extraction of mandarine peels have significant benefits over classic extraction and extraction with supercritical carbon dioxide. Observing total polyphenols values there is significant increase (0.1 mg GAE/g for SCO2, 1.82 for conventional and 2.32 using ultrasonics). DPPH of ultrasonically obtained extracts was also significantly higher (7.48 mM Trolox/g compared to conventional extracts 5.66 mM). Using ultrasound in this setting can ensure efficient reuse of waste from fruit and vegetable products for the production of bioactive compounds.

Topics:
High Pressure
Food (Sustainability & Environment)
Food-Energy-Water Nexus

AUTHOR INDEX

AIChE
120 Wall Street, FL 23
New York, NY 10005-4020

ISBN 978-1-7138-0579-3

9 781713 805793